I0056740

Fundamentals of AI for Young Minds

A Fun and Easy Guide to Learning AI

By

Dr. Edward K. S. Buckman, PhD, MBCI, CBCP, ITIL

Illustrations and Learning Design by

Nexora AI Learning Lab

Published by

Nexora Press

First Edition – 2025

Welcome to the World of AI!

(For Curious Kids Ages 7 to 12)

Have you ever asked a smart speaker a question? Played a game that got better the more you played? Or seen a robot in a movie and wondered, "How does it work?" That's the magic of Artificial Intelligence, or AI—and it's becoming a big part of our everyday world.

This book, *Fundamentals of AI for Young Minds*, is your first step into understanding how AI thinks, learns, and helps people do amazing things. Whether you're a tech explorer, a future inventor, or just super curious, this book is made for YOU.

Inside, you'll discover:

- What AI means.

- How machines can learn like humans.

- Where AI shows up in your world (even when you don't notice!).

- What jobs AI can help with—and what it can't.

We'll explore these ideas together with fun pictures, simple explanations, and a helpful robot guide to make learning easy and exciting.

This book is part of the Nexora Press AI Learning Series, a collection created to help young minds like yours grow smarter with technology, one fun chapter at a time!

Are you ready? Let's dive into the future of AI—together!

Publishing Information

Title: *Fundamentals of AI for Young Minds*
Author: Dr. Edward Kofi Sasa Buckman
Publisher: Nexora Press
Year of Publication: 2025
ISBN: 979-8-9991843-0-6
Edition: First Edition
Version: 1.0
Printing Location: Printed in the United States of America
Publisher Contact:
Nexora Press
404 South Roan Street
Johnson City, TN 37601
Email: ebuckman@nexorapress.com
Website: www.nexorapress.com

Legal & Copyright Information

Dedication

To my beloved wife,
whose unwavering support, love, and encouragement
continue to inspire every step of this journey.

And to my family,
whose presence and belief in my vision
give meaning to every word written on these pages.

This book is for you.

Preface

Welcome to *Fundamentals of AI for Young Minds*, a fun, engaging journey into the world of Artificial Intelligence, created especially for children and the adults who guide them.

As AI becomes part of our daily lives—from talking speakers to smart apps and helpful robots, it's more important than ever that young minds use this technology and understand it. This book aims to explain AI by breaking it down into simple concepts children can grasp, using clear explanations, fun illustrations, and real-life examples they see around them.

This book is more than just facts—it's an invitation to explore, imagine, and ask questions. Each chapter includes activities, discussions, and quizzes to make learning interactive and enjoyable. Whether you're a teacher, a parent, or a young explorer flipping through these pages, you'll find plenty of "aha!" moments along the way.

Fundamentals of AI for Young Minds is the first in a growing series to guide children from basic AI awareness to more advanced understanding as they grow. We aim to inspire the next generation of thinkers, problem solvers, and creators who will shape the future with heart, ethics, and imagination.

Let's start the journey together.

The Author

Dr. Edward Kofi Sasa Buckman is an educator, technologist, and passionate advocate for early AI literacy. With a background in cybersecurity and artificial intelligence, Dr. Buckman has worked extensively in academia and industry, helping people understand the evolving relationship between technology and society.

Inspired by a deep belief that the future begins in today's classrooms, Dr. Buckman created *Fundamentals of AI for Young Minds* to introduce children to the exciting world of smart machines, ethical thinking, and responsible innovation. My mission is to make technological education accessible, fun, and inspiring for learners of all ages.

Outside of writing and research, Dr. Buckman enjoys mentoring young students, exploring emerging tech trends, and spending quality time with family. This is the first book in a series dedicated to nurturing the next generation of curious, creative problem solvers.

Table of Contents

Introduction

Have you ever talked to a robot, asked a question to a voice assistant like Siri or Alexa, or seen a smart vacuum clean the floor by itself?

If so, you've already met Artificial Intelligence—AI for short!

Meet Nora, your AI guide!

This book is here to answer some big questions in a way that's easy to understand:

- What is AI?

- How does AI learn?

- Can AI think like a person?

- And how is it changing the world we live in?

You'll meet friendly robots, explore excellent AI tools, and discover how machines can "see," "hear," and "think" in their own way. Along the way, you'll also learn how to stay smart and safe when using technology.

Each chapter contains stories, pictures, games, and cool facts designed for YOU. And don't worry—no hard math or computer science required!

So, are you ready to become an AI Explorer?
Turn the page and let's begin!

Chapter 1
What is AI?

Chapter 1: What Is AI?

Learning Goals:

- Understand what AI means.

- Recognize AI in everyday life.

- Spark curiosity about how machines "think".

Activity Time!

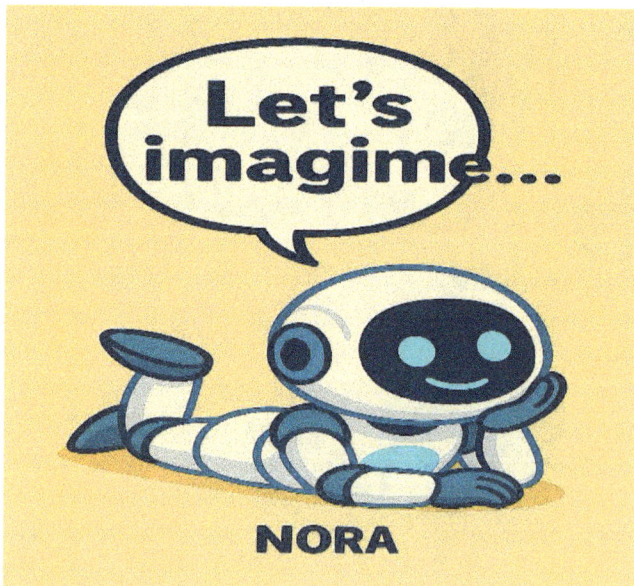

Let's Imagine...

Have you ever talked to Alexa or Siri? Did you watch YouTube and get the perfect video suggestion? Or played a video game where the computer plays against you? If you said "yes," you've already met AI!

What is AI?

AI stands for **Artificial Intelligence**.

That's a fancy way of saying:

"Machines that can think, learn, and solve problems like people do!"

But wait — can machines think? Not like we do! But they can learn patterns, make decisions, and **improve over time**.

AI All Around Us

Here are some places you might see AI in action:

AI Tool	What It Does
YouTube	Recommend videos you might like.
Smart speaker (like Alexa)	Answer your questions.
Voice typing	Type what you say.
Google Maps	Shows the fastest way to go.
Video games	Makes the computer player smart.
TikTok or Snapchat	Adds filters to your face in real-time.

Why Is AI Important?

AI helps make life easier and faster! It helps:

- **Doctors** find diseases.

- **Cars** drive themselves.

6

- **Robot cleans** floors.

- **Apps** understand what we say.

- Pretty cool, right?

Activity Time!

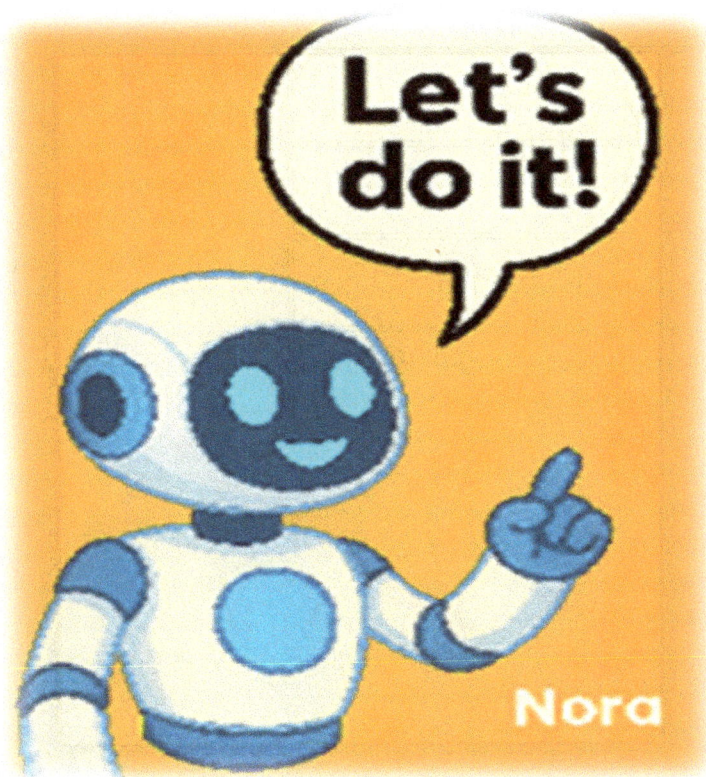

AI Around Me

1. **Activity**: List five things you used today that might use AI. Then, circle the one you use most often.
2. **Draw It**: Draw a robot doing something you'd like to help with (like homework, cleaning your room, or cooking)!

New Knowledge Recap

- **AI** = Artificial Intelligence

- AI is in phones, games, apps, and even cars!

- AI helps machines learn and make smart choices

Key Words

Word	What It Means
AI	Artificial Intelligence — smart machines!
Robot	A machine that can do things on its own.
Learn	Getting better by practicing and using data.

Mini Quiz (Just for Fun!)

1. What does AI stand for?

   - a) Amazing Ice-cream 🍦

   - b) Artificial Intelligence 🤖

8

◦ c) Animal Island 🏝️

2. Is Siri an AI?

◦ Yes / No

3. Name one way you used AI today: _____

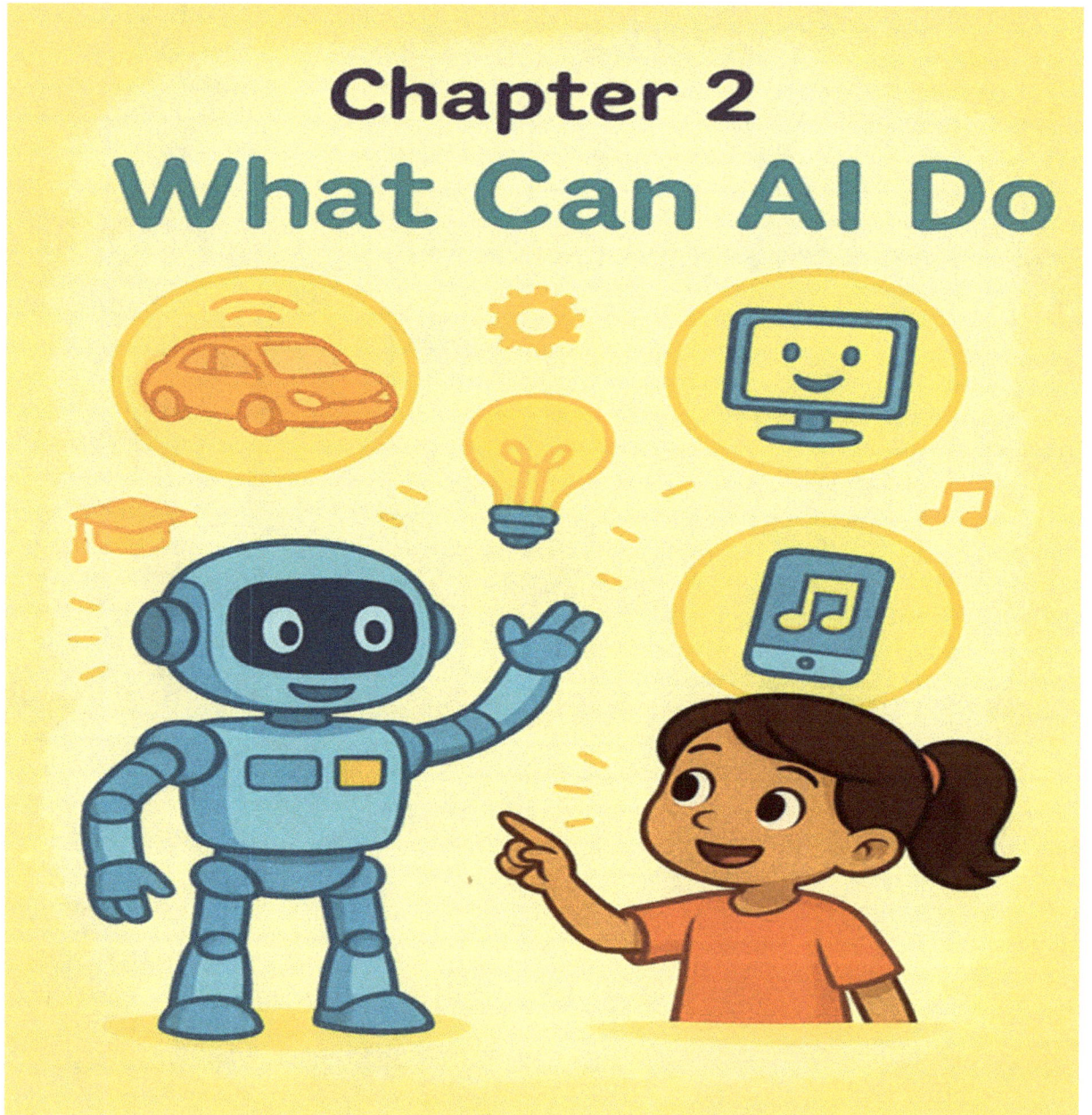

Chapter 2: What Can AI Do?

Have you ever asked a smart speaker to play your favorite song? Or watched a movie that was *just right* for you? That's AI at work!

Artificial intelligence (AI) can help us in many ways. Let's meet some of them!

AI Around You

AI Can See

AI can look at pictures or videos and recognize things, like cats, cars, or even faces!
"Hello! That's a stop sign," says the self-driving car's camera.

AI Can Listen and Talk

AI can hear your voice and talk back.
Like when you say, "What's the weather?" and it answers, "It's sunny today!"

AI Can Move

Some AI lives in **robots** that can move, dance, or even clean the floor!
"Zzzzz... I'll clean up your crumbs," says the robot vacuum.

AI Can Choose or Recommend

AI helps us find new music, play fun games, or show us what we might like.
"You liked that game, so you might like this one too!"

AI Can Count, Predict, and Help Plan

AI can analyze a large amount of information and make guesses, such as the weather tomorrow or the length of a trip.

When you use a search engine, talk to a chatbot, or play with a robot toy, AI helps make it work!

Little Thinker Box

Can you spot something in your house that might use AI?
Draw it on paper or discuss it with your class or family!

Key Words

Word	What It Means
Task	A job or thing that needs to be done.
Recognize	To know or understand something when you see it.
Smart	Able to think, learn, or solve problems well.
Pattern	Something that repeats or follows a rule.

Mini Quiz (Just for Fun!)

1. Which of these can AI **NOT** do (yet)?

    o a) Play chess

    o b) Write stories

    o c) Hug you like a friend

2. AI can see and understand pictures.

    True or False _____.

3. Fill in the blank:

    AI is good at tasks with clear _____.

Chapter 3
How AI Learns

Chapter 3: How AI Learns

Learning Goals:

- Understand that AI learns from data.

- Discover how AI finds patterns.

- Explore the difference between learning like a human and like a machine.

Can a Robot Go to School?

Imagine if a robot went to your school. It would sit in class, listen to the teacher, and do homework. But... would it learn like you do?

Not quite.

You learn by **thinking**, **asking questions**, and **trying new things**. AI learns differently — by looking at **lots and lots of data**.

What is Data?

Data is information.
It can be:

- Pictures

- Words

- Numbers

- Sounds

AI needs **a lot of data** to learn something.

Example: To teach an AI what a cat looks like, show it **thousands of cat pictures**. It starts to notice: "Hmm... small ears, whiskers, soft fur. Got it!"

Finding Patterns

AI learns by finding **patterns**.

Just like you might notice:

- Ice cream melts when it's hot.

- Dogs bark when they're excited.

- Your best friend always picks chocolate milk at lunch.

AI spots patterns in the same way, but faster and with **more examples**.

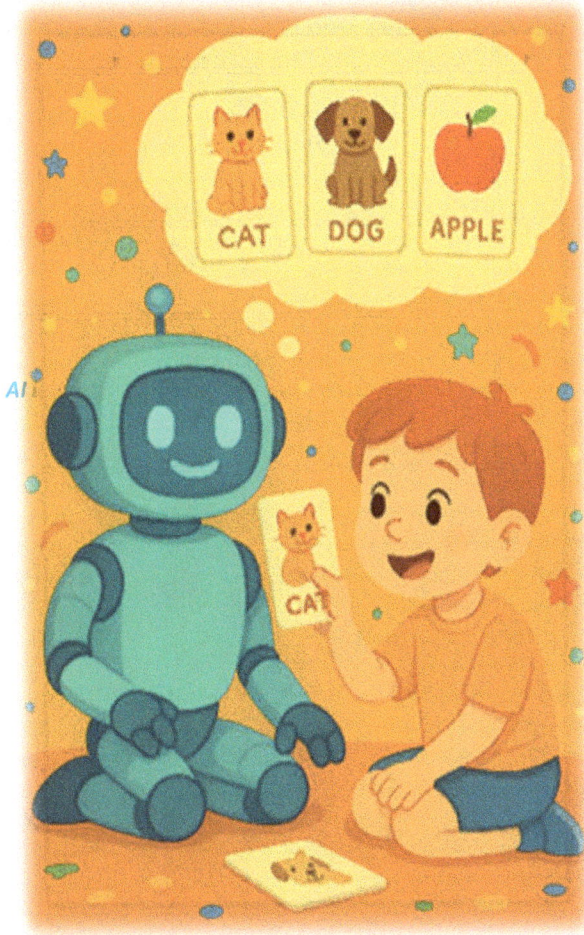

AI vs. Human Learning

Human Brain	AI Brain
Learns from feelings, mistakes, and experience.	Learns from data and examples.
Can learn from just one try.	Needs lots of examples.
Understands jokes and emotions.	Doesn't really "have" feelings.

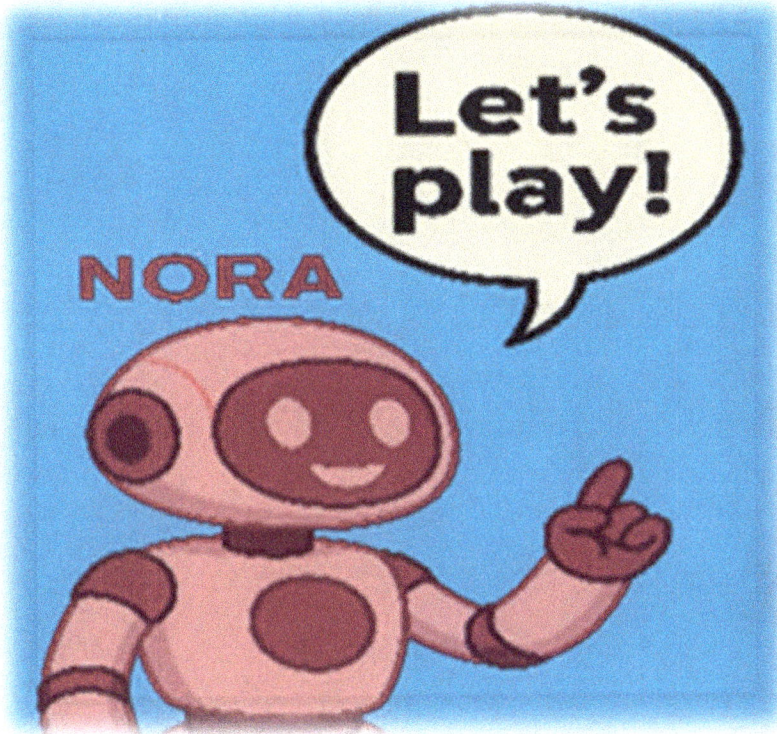

Pattern Detective

Look at the pictures below:

🍎 🍌 🍎 🍌 ?

What comes next? If you said 🍎 , you found the pattern!

That's how AI tries to guess what comes next by spotting patterns.

Teach an AI!

Try this online activity with help from an adult or teacher:

- Go to

  teachablemachine.withgoogle.com

- Use your webcam to show the AI

  pictures of:

  ○ You with a smile 🤪

  ○ You with a frown 😡

- Train the machine

- See how it learns to tell the

  difference!

That's **machine learning** in action!

New Knowledge Recap

- AI learns using **data.**

- It looks for **patterns** in pictures, words, and numbers.

- Humans and AI learn differently, but both can be smart in their own way.

Key Words

Word	What It Means
Data	Information (like pictures, numbers, or words)
Pattern	Something that repeats or follows a rule
Learn	Getting better by practicing or studying

Mini Quiz

1. What does AI need to learn?

- ○ a) Toys 🧸

- ○ b) Data 📊

- ○ c) Chocolate 🍫

2. What is a pattern? _____.

3. True or False: AI learns just like humans _____.

Chapter 4
Where We See AI

Chapter 4: Where We See AI

Learning Goals:

- Identify real-life examples of AI.

- Understand how AI helps in homes, schools, hospitals, and games.

- Get curious about how AI changes the world around us.

Is AI Only in Robots?

When people hear "AI," they often think of walking, talking robots from movies.

But guess what?

AI is already all around us, and it's usually invisible! You've probably used it today—maybe even before breakfast! Let's explore where we see AI every day.

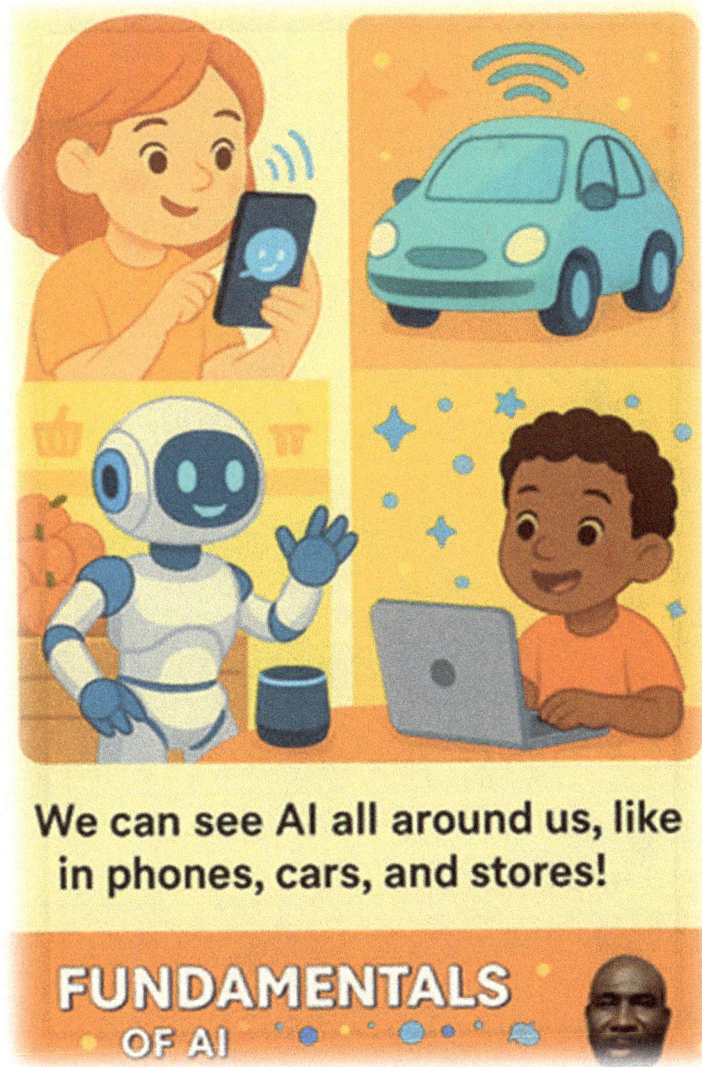

We can see AI all around us, like in phones, cars, and stores!

FUNDAMENTALS
OF AI

At Home

- **Smart Speakers** (like Alexa or Google Assistant)
→ "Hey Google, what's the weather today?"

- **Smart TVs**
→ Suggests shows you might like.

- **Smart Thermostats**
→ Changes the temperature based on what you like.

- **Robot Vacuums**
→ Cleans the floor by itself!

At School

- **Typing Suggestions**
  → Google Docs guesses what you're going to type next.

- **Translation Tools**
  → Helps you understand different languages.

- **Learning Apps** (like Duolingo or Khan Academy)
  → Gives you questions based on what you know.

In Hospitals

- **AI Scans X-rays**
  → Helps doctors spot problems faster.

- **Smart Watches**
  → Monitors your heartbeat and health.

- **Robot Helpers**
  → Bring tools to the doctor's surgery.

On the Road

- **Self-Driving Cars**
  → Drive using cameras and sensors.

- **Traffic Apps** (like Google Maps)
  → Finds the fastest route.

In Fun and Games

- **Video Games**
  → The computer players learn how to beat you!

- **Music Apps**
  → Suggest songs you might love 🎵

- **Face Filters**
  → Make your face look like a dog 🐶 or superhero 🦸

Play Time

AI or Not AI?

Circle the ones that use AI:

- a) A chair 🪑
- b) Google Maps 🗺️
- c) A toaster 🍞
- d) Netflix 🍿
- e) A fridge with a touchscreen 🧊
- f) A paper notebook 📓

AI Poster Time!

Create a poster called

"AI Around Me!"

Draw 3–5 things you use every day that might use AI.

Add a fun title and color it.

Share it in your class or with your family!

31

New Knowledge Recap

- AI is in many places: homes, schools, hospitals, and games.

- Some AIs are visible (like robots), and some are hidden (like inside apps)

- You use AI more often than you think!

Key Words

Word	What It Means
Smart Device	A gadget that can learn or make decisions
App	A program that does something useful or fun
Suggestion	A helpful guess or idea

Mini Quiz

1. Where might AI help you learn a new language? _____.

2. Which of these uses AI?

    o a) Traffic lights

    o b) Google Maps

    o c) A ball

3. Draw a place where you think AI is helping.

Chapter 5
The Logic Behind AI

Chapter 5: The Logic Behind AI

Learning Goals:

- Understand how AI follows rules and logic

- Learn about "If... then..." thinking

- Try fun activities that show how AI makes decisions

Can AI Think for Itself?

AI doesn't *really* think like we do.

It doesn't get happy, sad, or hungry 🍕 .

Instead, it follows **rules**. These rules help it decide what to do in different situations.

What Is Logic?

Logic is a way of thinking that follows clear steps.

AI uses logical steps to make decisions.

Example:
If it's raining, then I'll take an umbrella. ☂️
If it's sunny, then I'll wear sunglasses. 🕶️

That's **logical thinking** — and AI uses it all the time!

AI Uses Rules Like:

Situation	AI Rule
If you click "play" on a video	Then start the video
If you type "cat" into a search	Then show cat pictures
If it sees a stop sign	Then tell the car to stop 🚗

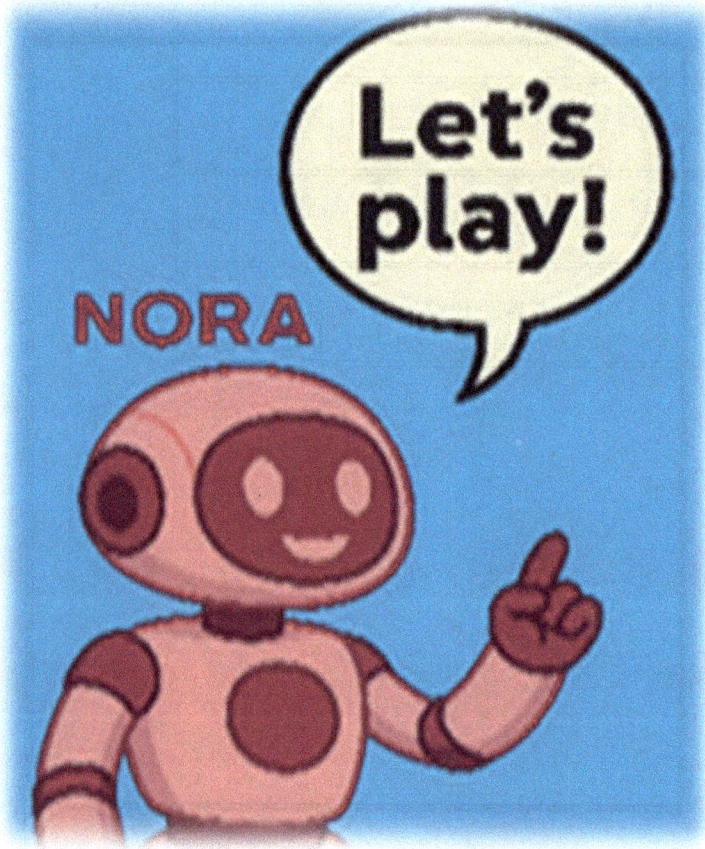

Mini Game: Robot Brain

Pretend you're a robot! Follow these rules:

1. If someone says "hello," then wave your hand. 👋

2. If someone says "bye," then spin in a circle. 🔄

3. If you hear a clap, then sit down. 🪑

Have a friend test you with these commands!

This is how AI **follows instructions** based on "if... then..." logic.

Coding and AI

When people build AI, they often use something called **coding**.

Coding tells computers what to do using step-by-step commands.

You can learn coding with block games like Scratch or Code.org!

"Want to become a coding pro? Check out our Coding for Young Minds book series — it's the perfect way to start your journey into the world of code!"

New Knowledge Recap

- AI doesn't feel—it follows rules and logic.

- It uses "if... then..." to make decisions.

- AI logic helps it act smart in many situations.

Key Words

Word	What It Means
Logic	Clear steps to solve a problem or make a choice
Rule	A command that tells what to do
If... Then...	A way of making decisions ("If this happens, then do that")

Mini Quiz

1. What does AI use to make decisions?

   - a) Feelings

   - b) Rules

   - c) Guesswork

2. What will a robot do if its rule says:
   If someone claps, then sit down?

   ➤ What happens when you clap? _____.

3. Write your own AI rule!

➤ *If _____, then _____.*

Chapter 6
Can AI Be Creative?

Chapter 6: Can AI Be Creative?

Learning Goals:

- Understand how AI makes music, pictures, and stories.

- Explore the idea of creativity in humans vs. machines.

- Try creative activities using AI tools.

What Is Creativity?

Creativity is making something new and special. It can be:

- A drawing 🎨

- A story 📚

- A dance 💃

- A new idea 💡

Humans are born with creativity. But can **AI** be creative too?

Can a Robot Paint?

Surprise! AI can:

- Draw pictures 🎨

- Write songs 🎵

- Create funny poems 🤖 ✍️

- Even design video game levels! 🎮

But wait... is that *real* creativity?

How Does AI Make Art?

AI doesn't imagine things the way we do.
Instead, it learns from **millions of pictures, sounds, or stories** and
mixes those ideas to create something new.

It's like:

Learning from every book in the library,
then writing your own story using parts of them!

Human vs. AI Creativity

Human	AI
Has feelings and dreams	Has data and rules
Can create from imagination	Combines what it has learned
Makes mistakes in new ways	Follows patterns in data

So, is AI creative? It depends on what you call creativity!

AI is smart, but it doesn't feel inspired like we do.

Let's Do It: Try AI Art!

Want to see what AI can create? Try this with help from an adult:
Activity Time!

1. Go to Craiyon.com or Google's AI Music. Type something fun, like:

   o "A robot eating spaghetti" 🍝

   o "A unicorn flying through space" 🦄 🚀

2. Watch as AI draws it for you!

Try changing the words and see what happens!

Activity Time!

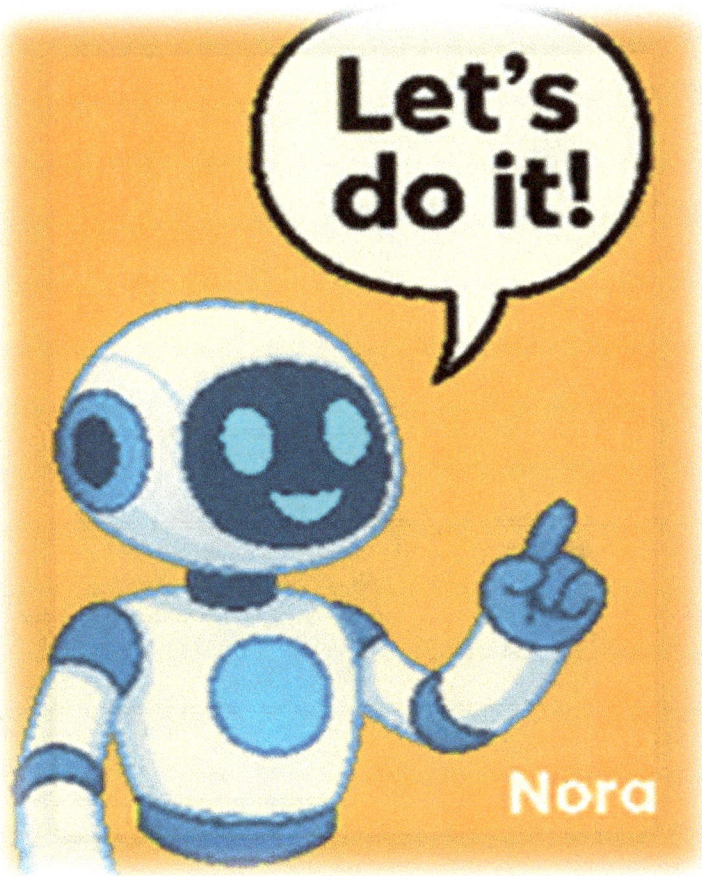

Let's Create Together

Here's a fun challenge:

- Draw a picture.
- Ask the AI to draw the *same picture*.
- Compare your art and the AI's art. Which one feels more fun, wilder, or more emotional?

New Knowledge Recap

- AI can draw, write, and even make music.

- It learns from examples and mixes them.

- People create with feelings—AI creates with data.

Key Words

Word	What It Means
Creative	Making something new and special
Art	A way to show ideas with pictures, sounds, or movement
Imagine	To picture something in your mind

Mini Quiz

1. What does AI use to make creative things?

    ○ a) Dreams

    ○ b) Data

    ◦ c) Magic

2. What's one thing AI can create? _____.

3. Draw your own creative idea, then ask AI to make one too!

CHAPTER 7
Can AI Be Right or Wrong?

Chapter 7: Can AI Be Right or Wrong?

Learning Goals:

- Understand that AI can make mistakes.

- Learn how AI can be unfair or biased.

- Discover why humans are essential for checking AI's decisions.

Does AI Always Get It Right?

AI can be super smart, but it's **not perfect**.

Sometimes, AI makes **wrong guesses**.

Can AI be right

Example: An AI is supposed to spot apples 🍎 , but it calls a red ball ⚽ an apple. Oops!

Why does that happen?

Garbage In, Garbage Out

There's a saying in computer science:
"Garbage in, garbage out."

This means:

- If you give AI **insufficient data**, it will give you **bad answers**.

Example: If you only show an AI dog pictures of **white poodles**, it might think black labs aren't dogs!

What Is Bias?

Bias is when something is unfair.

An AI is **biased** if it treats people unfairly.

😀 😀 😀

If an AI only sees faces of one skin color, it might not recognize others. That's bias.

AI doesn't mean to be unfair, but if the **data is unfair**, the AI will be too.

Who Checks the AI?

Humans do!

People are needed to:

- Pick the correct data.

- Test the AI.

- Make sure it's fair to everyone.

AI needs human help to be responsible and kind.

53

Can AI Hurt People?

Sometimes AI is used in **important decisions**, like:

- Who gets a loan 💰

- Who gets a job 👩🏽‍🎓

- Who gets into school 🎓

If the AI is unfair or wrong, **honest people can get hurt.** That's why fairness matters!

Activity Time!

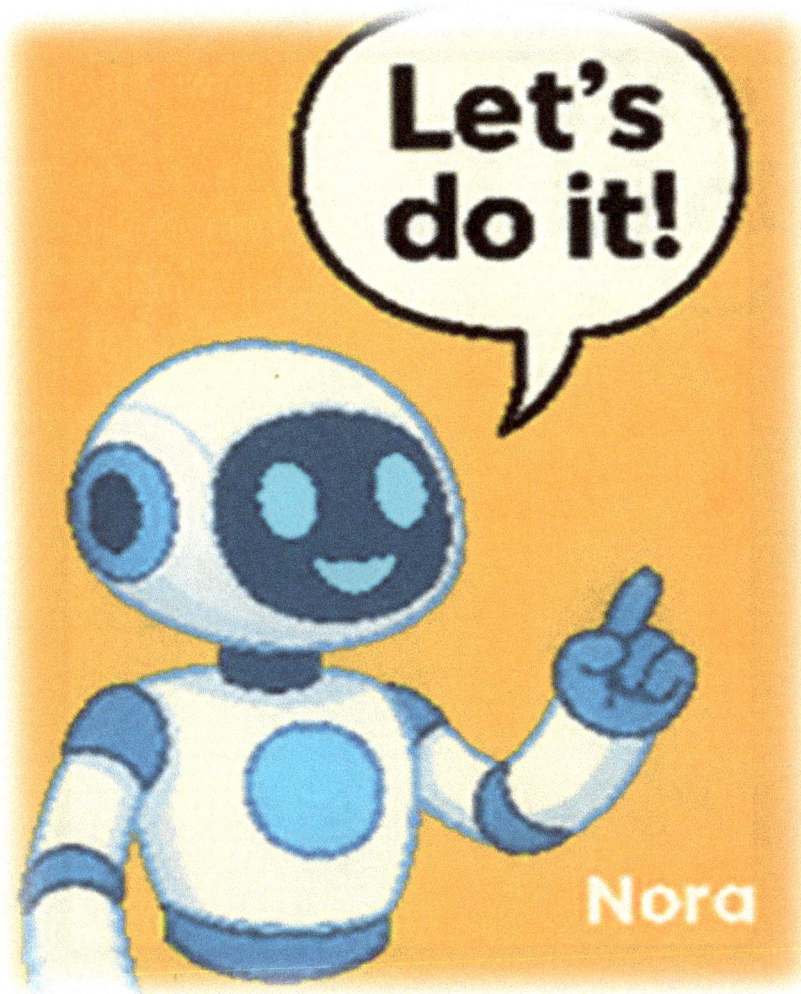

Let's Try: Spot the AI Mistake!

Look at this example:

You ask an AI: "Show me pictures of great scientists."

If the AI only shows pictures of **men**, is that fair?

No! Scientists can be men **or** women. That AI needs better data.

55

Let's Talk About It

Ask a grown-up or teacher:

- Can you think of a time when something was unfair?

- How could we teach AI to be fair?

Talking about fairness helps us teach AI to do better.

New Knowledge Recap

- AI can make mistakes.

- AI can be unfair if the data is unfair.

- Humans must check AI to make sure it's doing the right thing.

Key Words

Word	What It Means
Mistake	When something is wrong or not true
Bias	When something is unfair to a group of people
Fairness	Treating everyone equally and kindly

Mini Quiz

1. What happens if AI learns from unfair data? _____.

2. Who helps AI be fairer? _____.

3. True or False: AI is always right. _____.

CHAPTER 8
Jobs AI Can Do (and Can't)

Chapter 8: Jobs AI Can Do (and Can't)

Learning Goals:

- Discover how AI is used in different jobs.

- Understand the kinds of tasks AI is good at.

- Learn what jobs still need humans.

What Can AI Help With?

AI is now a helper in many workplaces! Let's explore where it shows up.

AI can do some jobs, but not others.

In Hospitals

- Helps doctors read X-rays.

- Reminds nurses when to give medicine.

- Tracks patients' heartbeats and steps.

But it can't:

- Comfort a patient 😊.

- Make tough decisions like a doctor.

In Banks

- Checks for fraud 🔍.

- Approves loans using numbers and data.

- Answers questions in a chatbot.

But it can't:

- Understand feelings if someone's sad or scared.

- Give personalized advice like a human banker.

In Cars

- Helps with self-driving.

- Parks the car for you.

- Finds the best route on maps.

But it can't:

- Handle surprises like a person on a bike suddenly appearing 🚴.

61

- Drive in crazy weather as well as a human.

In Restaurants

- Takes your order at a kiosk 🖥️.

- Suggests what you might like.

- Helps chefs with timing food.

But it can't:

- Smile and greet you.

- Handle special requests with care.

What Jobs Can AI Do?

AI Is Good At...	AI Is Not Good At...
Repeating the same task over and over 🌀	Being creative 🎨

AI Is Good At...	AI Is Not Good At...
Solving math or logic problems quickly 🧮	Comforting someone 🥹
Organizing big data 📊	Making moral choices ⚖️
Following clear rules 📋	Leading a team 🧍🧍

Why Humans Still Matter

AI can be **fast**, but **humans** can be:

- Kind ❤️

- Creative 💡

- Flexible 🤸

- Wise 👵

63

AI is a helper, not a replacement.

Activity Time!

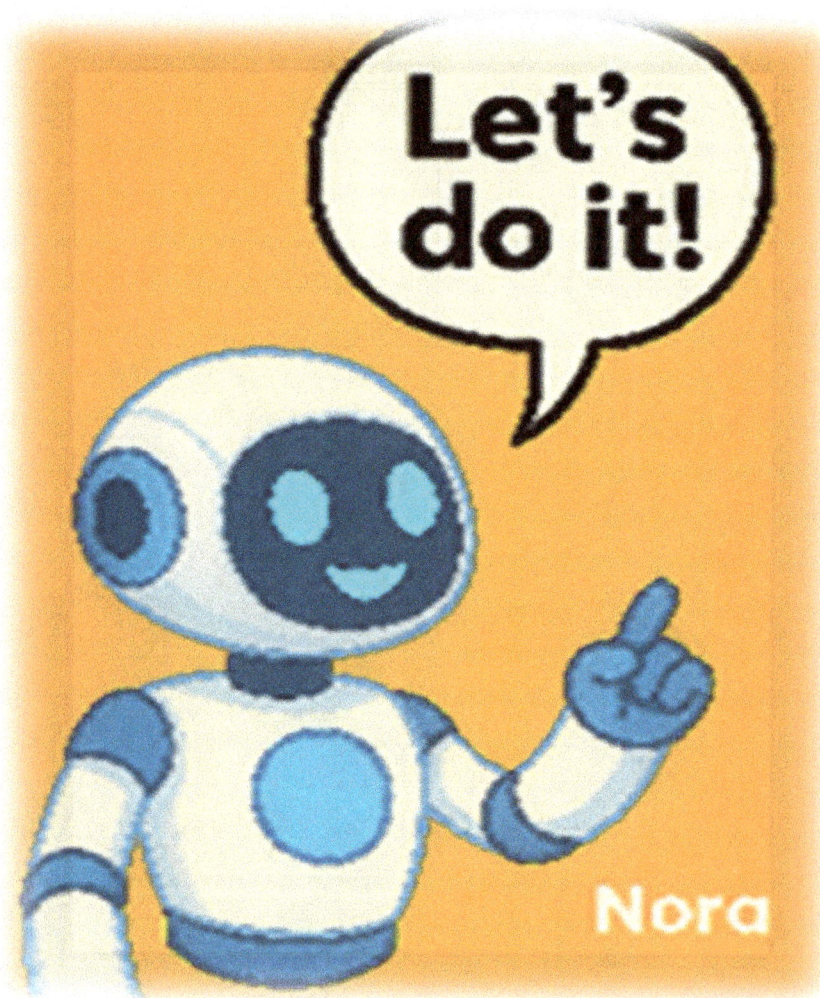

Job Sort!

Grab a paper and draw two columns:

1.   "AI Can Do   This"

2.   "Humans Can Do This"

Then list or draw some jobs, like:

- Baking a cake 🎂.

- Helping someone who is sick 🥺

- Driving a bus 🚌

- Writing a poem 📝.

- Counting money 💵

Sort them into the right columns!

New Knowledge Recap

- AI helps with many jobs.

- Some work needs a **human heart.**

- The best teams are **people + AI** working together.

Key Words

Word	What It Means
Job	The work someone does
Helper	Someone or something that makes work easier
Replace	To take the place of something

Mini Quiz

1. What is one job AI is good at? _____.

2. What is one job humans do better? _____.

3. Do you want to work with AI when you grow up? Why?

   _____.

Chapter 9: How to Talk to AI

Learning Goals:

- Learn how to ask clear questions to AI

- Understand what AI can and can't answer

- Discover safe and smart ways to use AI

Talking to a Machine?

You really can "talk" to AI!

You can:

- Ask questions

- Give it tasks

- Even have fun conversations!

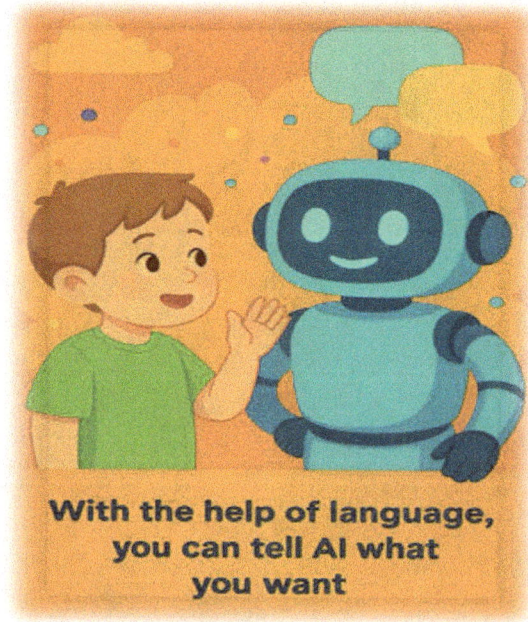

With the help of language, you can tell AI what you want

AI tools like ChatGPT, Alexa, or Siri are made to **listen and respond**—but not like real people.

What AI Understands

AI understands **clear instructions**.
If you say:

"Tell me a story about a dinosaur that bakes cookies" 🍪 🦖

It will try its best to make one!

But if you say something too confusing, like:

"Tell me about that thing with the stuff and the thingy..."

AI won't know what you mean.

How to Ask Good Questions

Not So Clear ❌	Better 👌
"Tell me stuff."	"Tell me 3 fun facts about space."
"I need help."	"Help me with my math homework on fractions."
"Say something."	"Can you tell a joke about a robot?"

Be specific. Be polite. Be safe.

AI Doesn't Know Everything

AI is **smart**, but:

- It can be **wrong** ❌

- It doesn't know the **future** 🔮

- It doesn't **have feelings** ❤️

So always double-check what it says, especially for school!

Be Safe When Using AI

Always follow these rules:

- 👍 Ask a trusted adult before using a new AI
- 👍 Don't share personal info (like your real name or address)
- 👍 Don't believe everything AI says
- 👍 If something feels weird, tell a grown-up

Play Time!

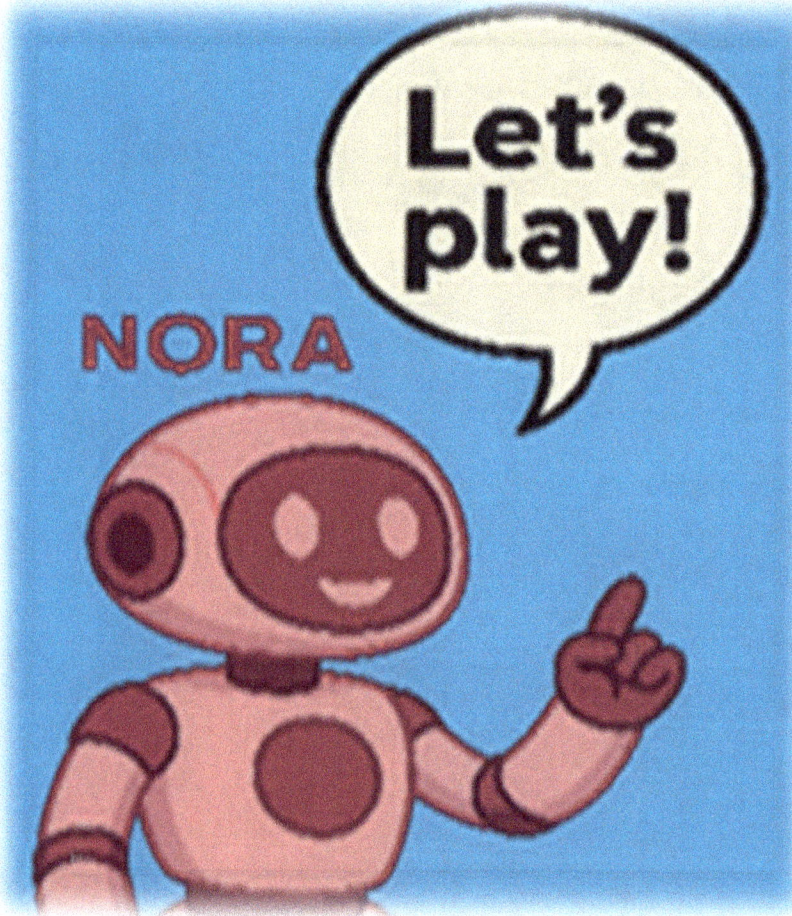

Try This: AI Question Game

1. Think of a question to ask an AI (like ChatGPT or Siri)

2. Ask it clearly

3. Read or listen to the answer

4. Ask a follow-up question

5. See how far the conversation goes!

New Knowledge Recap

- You can talk to AI if you use clear and safe questions

- AI gives smart answers—but it's not always right

- People must use AI wisely and kindly

Key Words

Word	What It Means
Question	Something you ask to get information
Safe	Protected from harm or danger
Respond	To answer back

Mini Quiz

1. What makes a question clear for AI? _____.

2. Should you tell an AI your real name or where you live? _____.

3. Write your own fun AI question to try later!

CHAPTER 10
HOW AI LEARNS
(AGAIN AND AGAIN!)

Chapter 10: How AI Learns (Again and Again!)

Learning Goals:

- Understand how AI learns from data

- See how practice helps AI get better

- Learn why AI needs feedback to improve

How Do You Learn?

Let's start with YOU!

You learn by:

- Listening 🎧

- Watching 👀

- Practicing 📝

- Making mistakes 😬

- Trying again 💪

AI learns the same way, but with data instead of feelings.

What Is Data?

Data is information. AI uses **lots and lots** of it.

If you show an AI **10,000 pictures of cats and dogs**, it starts to learn the difference between them. 🐱 🐶. Each picture is a **learning example**.

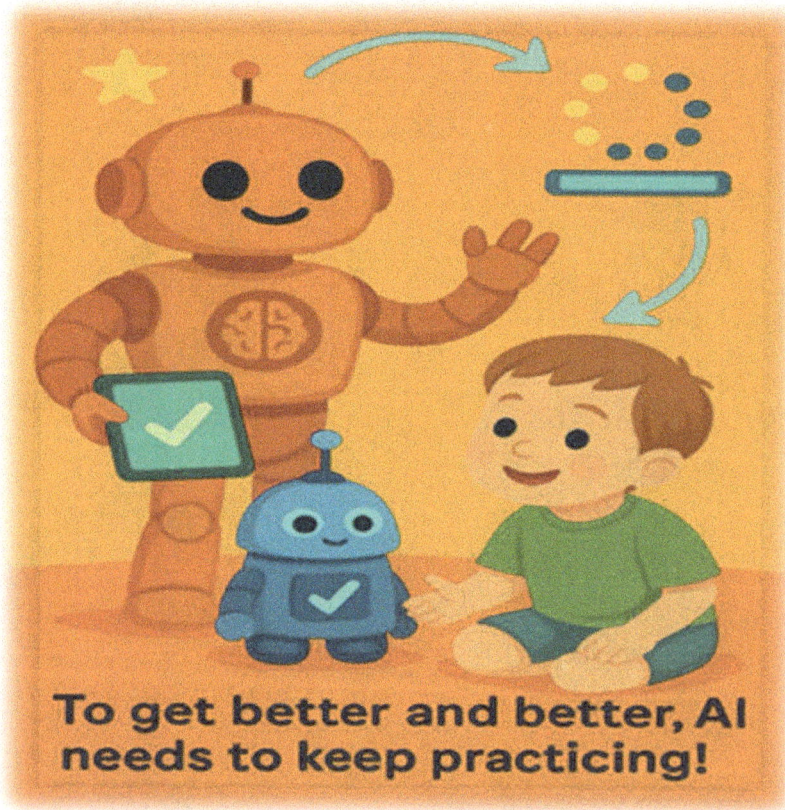

To get better and better, AI needs to keep practicing!

Practice Makes Progress

The more data AI sees, the better it gets.
But it has to **practice** again and again and again!

You get better at soccer ⚽ or spelling 🧠 the more you practice.

AI training can take hours, days, or even weeks!

Needs Feedback

AI gets feedback like this:

- If the answer is **right**, it gets a reward (like a point).

- If it gets it **wrong**, it adjusts and tries again.

This is called **machine learning**!

Think of it like a robot learning to sort apples and oranges.
At first, it makes mistakes, but feedback makes it smarter!

Can AI Learn the Wrong Thing?

Yes! If the data is:

- Wrong ❌

- Unfair 😣

- Too small 📉

Then AI might **learn badly**. That's why good data is super important.

Let's Try It: Teach Me a Pattern

Pretend YOU are the AI. Try this game with a friend:

1. Your friend claps a pattern like:

   ○ 👏👏🖐️👏👏🖐️

2. Your job: Try to **learn and copy the pattern.**

3. Keep practicing until you get it right!

That's how AI feels when learning from patterns, too!

AI Learns Every Day

Some AIs even learn from **new things people say**!

That's why you might notice AI getting better over time—it's always learning (with help from smart humans behind it!).

New Knowledge Recap

- AI learns from data, just like you learn from practice.
- Feedback helps AI improve.
- Good data = good learning!

Key Words

Word	What It Means
Data	Information used to learn
Feedback	Helpful response to improve learning

Word	What It Means
Pattern	Something that repeats in a certain way

Mini Quiz

1. What helps AI get smarter over time? _____.

2. What happens if AI gets bad data? _____.

3. What is something YOU had to practice to get better at? _____.

CHAPTER 11
WILL AI TAKE OVER THE WORLD?

Chapter 11: Will AI Take Over the World?

Learning Goals:

- Talk about fears and facts around AI.

- Understand what AI can and cannot do.

- Learn why humans are always important.

Big Question, Big Feelings

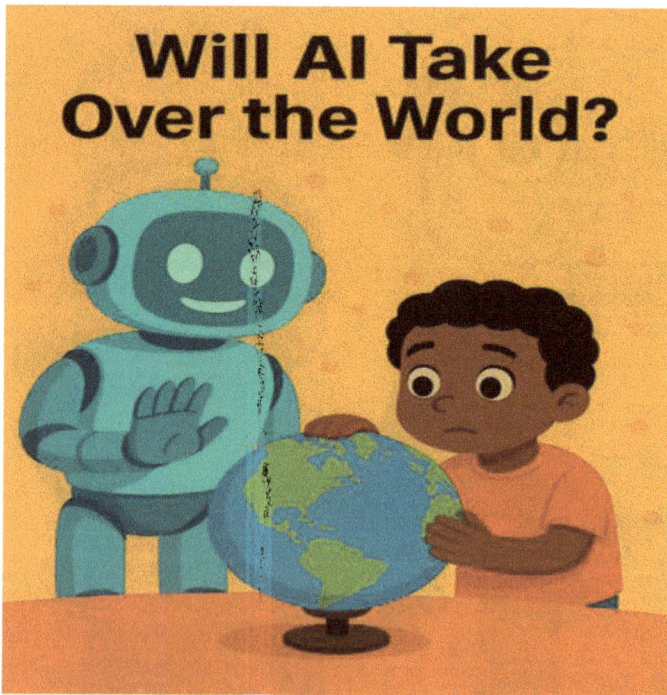

Some people worry that:

- "Robots will take all our jobs!"

- "AI will become smarter than humans!"

- "Machines will rule the world!"

Those sound like movie plots 🎬 ... but is that true?

Let's find out.

What AI Is

AI is **not magic** and **not a person**.

It's just a computer program that:

- Learns from data.

- Follows instructions.

- Helps us with tasks.

It **doesn't think or feel** the way humans do.

What AI Can't Do

AI can't:

- Make its own dreams 🌈

- Have feelings 🧡

- Decide what's right or wrong ⚖️

- Replace human love, kindness, or wisdom

It only works because **people built it** and tell it what to do.

Humans Are Still in Charge

Even the smartest AI:

- Needs humans to create it.

- Can be turned off 🔌

- Should be used with rules and care.

We call this **AI ethics**—making sure AI is safe, fair, and helpful.

AI Is a Tool, Not a Boss

Think of a:

- Hammer 🛠️

- Bicycle 🚲

- Calculator ➗

All tools that help us, but **don't take over**.

AI is the same! It's strong **with us**, not **against us**.

What About Sci-Fi Movies?

Robots taking over the world sounds exciting in movies. But in real life:

- AI doesn't want power.

- It doesn't make plans.

- It doesn't want anything at all!

Robots like WALL-E or evil computers in movies aren't real—they're made for fun.

Play Time!

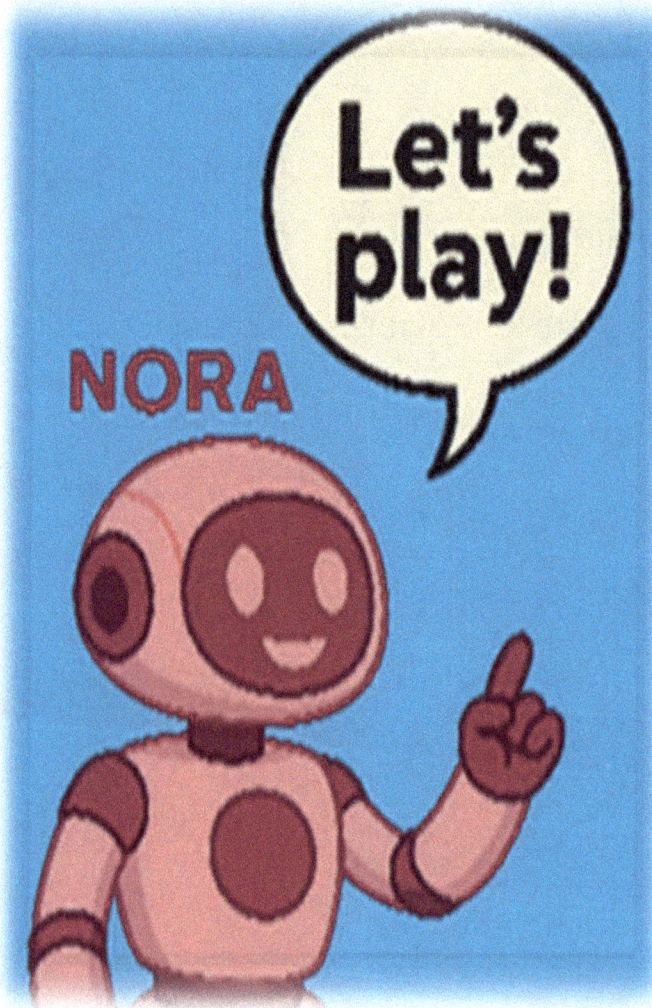

Let's Talk: Human or Machine?

Here's a fun thinking game:

Which of these needs a **human**?

• Hugging a friend 🤗

• Painting a new idea 🎨

• Solving a puzzle 🧩

• Making someone laugh 😂

Yes! All of them. AI can help, but only **people** bring their hearts and imaginations.

New Knowledge Recap

- AI is a tool made by people.

- It can't feel, dream, or decide.

- Humans guide how AI is used.

- Stories about AI "taking over" are fun, but not real.

Key Words

Word	What It Means
Ethics	Rules about what's right and wrong
Take Over	To control something completely
Tool	Something we use to help with work

Mini Quiz

1. Can AI think or feel like a human?

2. Who is in charge of how AI is used?

3. Have you seen a movie with robots? Was it real or pretend?

Chapter 12: Fun AI Projects You Can Try at Home

Learning Goals:

- Try simple AI-related activities.

- Understand how AI sees, hears, and learns.

- Build confidence by creating and exploring.

Ready to Be an AI Explorer?

You don't need fancy robots or a lab to play with AI.

Here are some fun and simple projects you can try at home, with help from your teacher or family.

1. Teach an AI to See (With Teachable Machine)

What You Need:

- A computer or tablet
- Internet access
- Visit: teachablemachine.withgoogle.com

What to Do:

- Use your webcam to train the AI to recognize different poses, toys, or faces.
- Show it a "thumbs up" 👍 and call it "Yes!"
- Show it a "thumbs down" 👎 and call it "No!"

- Then see if it can tell the difference!

You just trained a mini computer to "see" patterns!

2. Talk to a Voice Assistant

What You Need:

- A smartphone, Alexa, Siri, or Google Assistant.

Try Saying:

- "What's the weather like today?"

- "Tell me a robot joke!"

- "Can you play animal sounds?" 🐘 🐶 🐓

Watch what it says.

Then ask it something it might *not* know and see how it responds.

Tip: Notice how it works better with clear questions!

3. Design Your Own Robot Helper

What You Need:

- Paper, crayons, markers.

Draw a robot that:

- Helps with homework 📚
- Cleans your room 🧹
- Makes your favorite sandwich 🥪

Then give it a name and write one thing it can and **cannot** do!

Bonus: Label its sensors, arms, and speakers!

4. Try a Pattern Game (AI-Style)

AI loves patterns. Play a "guess the pattern" game with a friend:

- Clap, snap, stomp (repeat).
- Hide a secret rule like: "Every third sound is a clap."
- Can they figure it out?

This is what AI tries to do when it learns from data!

5. Try a Kid-Friendly Coding Site

Want to go a bit further? Try:

- <u>Scratch</u> – make games and animations.

- <u>Code.org</u> – explore AI with Minecraft or Star Wars.

- AI for Oceans – teach AI to spot trash vs sea creatures 🌊

Learning to code helps you speak the language of AI!

"Want to become a coding pro? Check out our Coding for Young Minds book series — it's the perfect way to start your journey into the world of code!"

New Knowledge Recap

- You can explore AI with fun, simple activities.

- AI learns from what we show and say.

- You're already on your way to being an AI inventor!

Key Words

Word	What It Means
Train	To teach something by showing examples
Pattern	Something that repeats
Project	A fun task you can build or try

Mini Quiz

1. What's one way to teach an AI using your webcam?

2. What kind of questions work best with voice assistants?

3. If you could build your own AI, what would it do?

Final Thoughts: You and the Future of AI!

Wow—you did it! You've just taken your first exciting steps into the world of Artificial Intelligence. From smart speakers to self-driving cars, from helpful robots to learning machines, AI is all around us— and now you understand how it works!

But here's the best part:
AI is still growing, and so are you.
The more curious you are, the more you'll discover. Maybe one day, *you* will build the next great AI that helps people, protects the planet, or solves big problems!

So keep asking questions.
Keep imagining new ideas.
And always remember—you don't have to be a grown-up to be a genius. The future needs young minds like yours.

Dream Big. Learn More. Create Tomorrow.

References

General Sources

1. **Teachable Machine by Google** – https://teachablemachine.withgoogle.com

2. **AI for Kids - MIT Media Lab Projects** – https://llk.media.mit.edu

3. **Common Sense Media AI Resources for Educators** – https://www.commonsense.org/education

4. **IBM Watson AI Education Materials** – https://www.ibm.com/watson-education

5. **Machine Learning for Kids by Dale Lane** – https://machinelearningforkids.co.uk

6. **World Economic Forum – Why Children Should Learn AI Early**

7. Simplified excerpts and inspirations from:

   - *"AI + Ethics Curriculum for Middle School"* – MIT & AI4K12.org

- *"Artificial Intelligence: A Guide for Thinking Humans"* by Melanie Mitchell (used indirectly for inspiration)

Author's Note

Dear Young Reader,

You've reached the end of *Fundamentals of AI* — and what a journey it's been!

From meeting robots and smart tools to learning how machines can "think," you've taken the first steps into the future. I hope this book helped you discover that AI is not just about technology — it's about **imagination**, **problem-solving**, and **curiosity**.

AI is shaping the world around us. And one day, maybe **you** will help build the next big thing!

Keep learning, keep exploring. The world of AI is just starting, and it needs young minds like yours.

With excitement for your journey,
Dr. Edward Kofi Sasa Buckman, PhD, MBCI, CBCP, ITIL
Educator | Technologist | AI Literacy Advocate

Acknowledgments

Special thanks to:

- Prince-William Ato Kwamena Sasa Buckman for proofreading the manuscript and encouraging the inclusion of more visuals to enhance learning.

- The countless educators, innovators, and researchers making AI accessible to all.

- Parents, teachers, and mentors who encourage young minds to explore technology.

- The young readers and future change makers who inspire us to create meaningful tools for learning.

www.ingramcontent.com/pod-product-compliance
Lightning Source LLC
Chambersburg PA
CBHW081821200326
41597CB00023B/4344